T3-BWQ-785

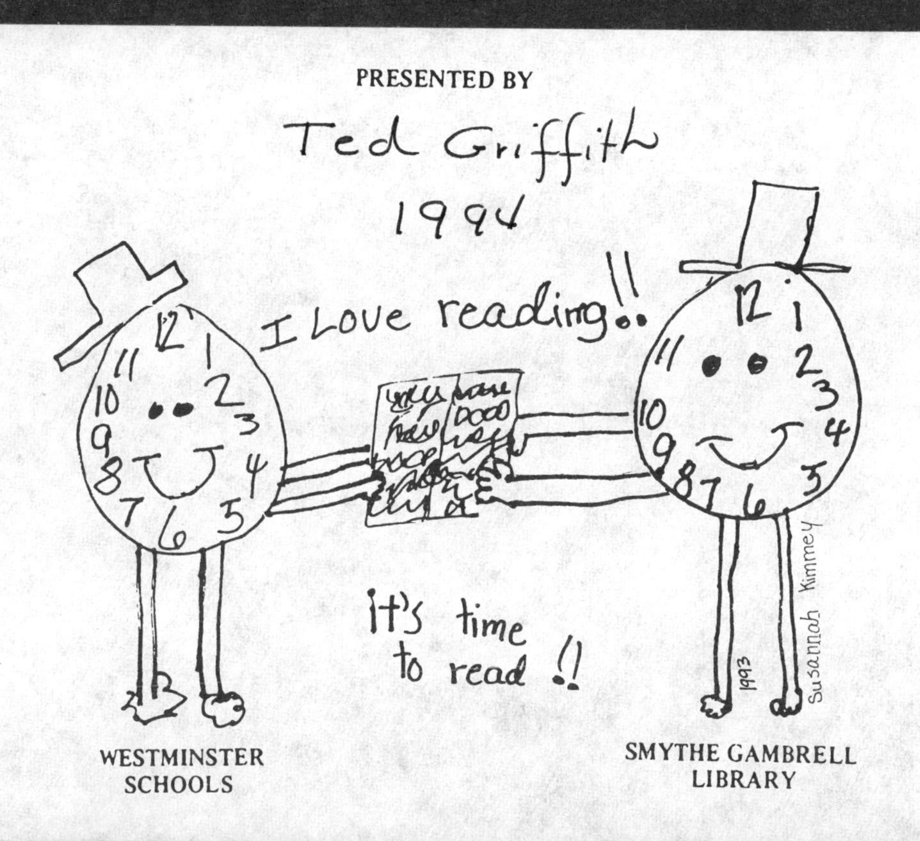
I Love reading!!
it's time
to read !!
1993
Susannah Kimmey

SPACE TRAVEL

FOR THE BEGINNER

SPACE TRAVEL FOR THE BEGINNER

Patrick Moore

Acknowledgements

My thanks are due to James Turner, aged (just) ten, who read through the manuscript of this book and made a number of useful comments and criticisms; I have followed his advice in every respect.

As on many previous occasions, I am most grateful to Paul Doherty for his splendid illustrations.

The title page picture is an artist's impression of Mars, by Paul Doherty.

The photographs were supplied by NASA and Patrick Moore

Published in North America by the Press Syndicate of the University of Cambridge
40 West 20th Street, New York, NY 10011-4211, USA
Published throughout the rest of the world by George Philip Ltd.
59 Grosvenor Street, London W1X 9DA

First published 1988
First Cambridge University Press edition 1992

Printed in Hong Kong

A catalog record for this book is available from the British Library

ISBN 0-521-41835-6 hardback

CONTENTS

CHAPTER 1
GOING INTO SPACE

Would you like to make a trip into space? Many people would say 'Yes'. It would be exciting; you would see and do things which would never be possible so long as you stayed on the surface of the Earth, and you would have no 'weight', so that you would float about inside your spaceship. Yet it would be dangerous, too; as we know, accidents have happened in the past.

The first spaceman was a Russian. His name was Yuri Gagarin, and he made his flight in April 1961. Since then there have been many other space travellers, more properly called 'astronauts', and 12 men have walked upon the surface of the Moon. New trips are being planned all the time, and it is quite likely that some of the readers of this book will themselves make journeys beyond the Earth. But as a start, let us see what has to be done, and where we can hope to go.

We live on the Earth, which is ball-shaped, and is moving round the Sun. The Sun itself is a star, and all the stars you can see on any clear night are themselves suns; they look much fainter and smaller than our Sun only because they are so much further away. (Stars cannot be seen in the daytime because the sky is too bright, just as you could not see the light of a match if you held it up in front of a powerful lamp.)

The distance between the Sun and the Earth is 93 million miles (150 million kilometres). This sounds a long way, but it is not much to an astronomer, who has to become used to talking about very great distances. If you could drive straight from the Earth to the Sun in a racing car, moving at a steady 100 miles (160 kilometres) per hour and never stopping, the journey would take you over 100 years – but it would take thousands of years to reach the nearest star.

The Sun is the centre of what is called the Solar System. ('Solar' comes from 'sol', the Latin word for 'Sun'.) Round the Sun move the nine planets, of which our Earth is one. Two of these planets are closer to the Sun than we are, while the others are further out. Some of the planets have smaller bodies or 'satellites' moving round them. The Earth has one satellite – the Moon – and we have always known that the Moon would be the first world which we would reach.

The Moon is only about a quarter of a million miles or 384,000 kilometres away. This is less than ten times the distance you would travel if you flew ten times round the Earth. A direct journey from the Earth to the Moon in our racing car would take less than four months.

In fact, things are not so easy as this. It is not possible to go from the Earth to the Moon (or any other world) by the shortest path. But the Moon

(*right*) The American *Saturn V* rocket stands on the launch pad at Cape Canaveral in Florida. This particular rocket was used for the *Apollo 17* Mission in December 1972.

(*below left*) Yuri Gagarin, the first man in Space, completed one orbit in the Soviet spacecraft *Vostok 1* on 12 April 1961.

seems to be within reach, and many years ago – long before the first rockets flew – stories were written about journeys there.

So far as I know, the first of these stories was written about 150 years after the time of Christ. In it, the author said that a ship carrying a full crew of sailors was caught in a waterspout, and thrown upwards so quickly that it landed on the Moon. This is great fun, but the author did not expect to be taken seriously; he even called his story the *True History* because it was made up of nothing but lies from beginning to end!

We have found out a great deal since then, but it is only during the past few years that we have really been able to send travellers into space. Even now, no men have been further than the Moon, though rockets which carry cameras have travelled out to the edge of the Solar System.

CHAPTER 2
THE EARTH AND ITS AIR

The fastest aircraft of today can travel at well over 1000 miles (1600 kilometres) per hour, but there is one very good reason why we can never use an aircraft to go to the Moon. This is because no aircraft can work unless there is air around it, and the Earth's air (better known as the atmosphere) does not stretch upward very far.

We live at the bottom of an ocean of air, and you can easily 'feel' it for yourself; simply cup your hand and wave it around, pushing the air out of the way. The higher up you go, the thinner the air becomes. Everest, the world's tallest mountain, is 29,000 feet (over 8800 metres) high, which is around 5 miles or 8 kilometres; at its top, the air is so thin that you cannot breathe it easily. If you go up to, say, 10 miles (16 kilometres), there is so little air left that you cannot breathe it at all, which is why all high-flying aircraft have to be carefully protected.

Earth from Space, as photographed by *Apollo 17*. Africa and Antarctica show clearly through our thin atmospheric layer.

Above 100 miles (160 kilometres) there is almost no air left, and above 1000 miles (1600 kilometres) it is safe to say that there is none at all. The Moon is more than 200 times as far away as that, so that most of the journey has to be done in empty space.

The old type of aircraft used to grip the air with a spinning propeller, forcing the air under its wings to keep itself up. Without air around it, the propeller has nothing to grip, and the aircraft is useless. The jet aircraft of today works in a different way, but it still needs air to suck into its engines, so that it too is useless above a height of a few tens of miles or kilometres. We must look for something quite different if we hope to travel into space.

One idea, put forward over a hundred years ago, was that of a 'space gun'. It was described in a famous story written by the French author Jules Verne (which is still worth reading, even though we now know that his ideas about space travel were wrong). In Verne's book, the travellers were put inside a hollow bullet and fired to the Moon at 7 miles (11 kilometres) per second, which works out at about 25,000 miles (40,000 kilometres) per hour. This would have meant reaching the Moon in less than three days.

Why did Verne choose this speed – 7 miles per second? The reason is that this is the Earth's 'escape velocity', a term which needs to be explained carefully, because it will be used over and over again all through this book.

If you take a solid object (a cricket ball will do) and throw it upward, it

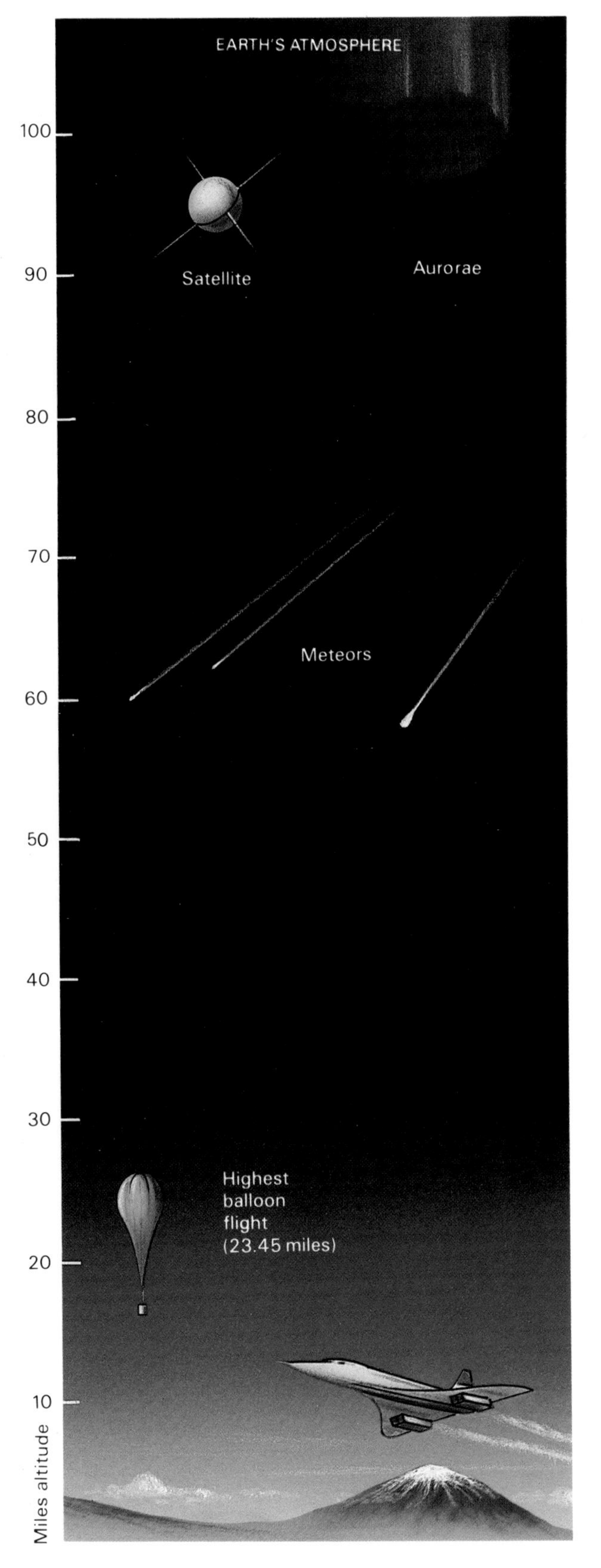

will rise to a certain height and then fall down. If you throw it faster, it will rise higher. The faster you throw it, the higher it will go. If you could throw it upward at 7 miles per second, it would never come down at all. The Earth's pull of gravity would be unable to hold it, and the ball would escape into space. (Try it if you like, but you will fail. The world's fastest pitchers cannot pitch a baseball even at 100 miles per hour!)

You can now see why Verne chose this speed for his bullet. If it were fired at less than escape velocity, it would simply travel upward for a time and then fall down. To get to the Moon, it had to start at the full escape velocity of 7 miles per second.

All this is true enough, but there are other difficulties with the space gun. If you rub against the air too quickly, you will set up heat by friction, because the air is being pushed out of the way and tries to stay where it is. You can show this quite easily every time you blow up the tire of a bicycle; the pump will become warm, because you are 'squashing' the air and causing friction. A bullet moving through the air at 7 miles per second would simply burn away. Quite apart from this, the shock of starting off at escape velocity would cause such a jerk that the space travellers would be turned into jelly.

Either Verne did not know this or else, for the sake of a good story, he did not bother about it. We know better, and we have found that there is only one way to go into space. This is by using the power of the rocket.

Earth's atmosphere becomes thinner the higher you go. Most artificial satellites orbit above 90 miles, where the air is virtually non-existent.

CHAPTER 3
THE ROCKET

Rockets were invented hundreds of years ago, and there can be few people who have not used them as fireworks. In England, they let them off every 5 November. It is worth finding out more about rockets, because display fireworks work on exactly the same principle as the space rocket.

The firework is made up of a hollow tube filled with gunpowder. At one end there is an exhaust, and a stick is added to make the rocket fly smoothly. On the case of the rocket you read: 'Light the blue touch-paper and stand well clear' – in other words, put your match to the paper and then jump clear! As soon as the gunpowder starts to burn, it gives off hot gas. This gas tries to get out of the tube, but it can do so in only one direction: where the touch-paper has been burned away. The gas rushes out through the exhaust, and in so doing it 'kicks' the tube in the opposite direction, so that the rocket flies.

The main point is that this will work whether or not there is any air around the rocket. It 'kicks against itself', so to speak, and air is actually a nuisance, because it sets up friction and has to be pushed out of the way.

It is easy to see what is meant without going to the trouble of buying a firework; a balloon will do. Blow the balloon up as hard as you can, holding the exhaust, and then suddenly let go. As the air inside rushes out, the balloon will shoot across the room. It will not go far, because there is not much gas inside it, and it will wriggle around; but if it were moving in space, it would move in a straight line. Once again, it moves because it is 'kicking against itself'.

If you are still not happy about this, put a piece of card on a smooth floor, tread on it, and then jump off to one side. You will find that the card will move in the opposite direction, simply because you have kicked against it. The same thing would happen even if there were no air in the room.

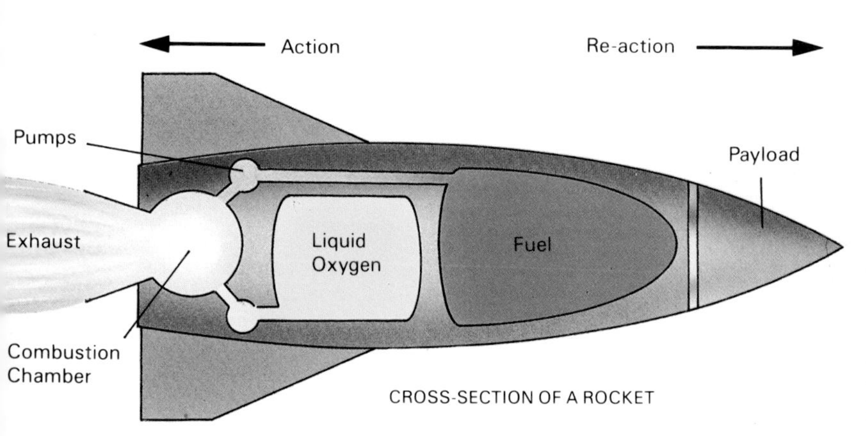

(*left*) Rockets are driven by the gas produced when the fuel and oxidant combine, causing spontaneous combustion.

(*right*) The Space Shuttle lifts off, using a combination of liquid fuel with solid fuel boosters.

The first man to understand this properly was the great English scientist Isaac Newton, over 300 years ago. It was also Newton who explained the laws of gravity. Gravity is the force which holds us down to the Earth; you feel heavy because the Earth is trying to pull you toward its centre. All bodies have their own gravity, and the

more massive the body the greater is its pull. The Sun is much more massive than the Earth, and has a much stronger pull; the Moon is much less massive, so its gravity is weaker. As we will see later, you would weigh less on the Moon than you do on the Earth.

Solid fuels such as gunpowder are much too weak to send a spaceship to the Moon, and a rocket motor of the sort used today has liquid fuels. There are two liquids of different types; they are pumped into the same 'chamber', as shown here, they will start to give off gas as soon as they are mixed. This gas then comes out through the exhaust, and the rocket flies.

A liquid-fuel motor of this type can be controlled, while gunpowder cannot. When you light the powder in our firework, it will burn until all of it has been used up, but with liquids you can alter the rate of mixing. This means that the rocket can start off slowly, and work up to the full escape speed only when it is above the thickest part of the Earth's air and is no longer in danger of being burned away by friction.

One trouble which spaceship planners have to face is that even liquid fuels are not as powerful as needed. The answer is to make 'step-rockets'. One rocket is mounted on top of the other, so that the large bottom rocket starts off the journey and then breaks away, falling back to the ground while the upper rocket continues the trip by using its own engines.

These ideas were put forward almost a hundred years ago by a Russian with the tongue-twisting name of Tsiolkovskii (pronounced 'Zee-ol-kov-ski'). He died in 1934, but it was not until 1957 that the Space Age really began.

CHAPTER 4

MAN-MADE MOONS

Many of the famous stories in history are not true. For example, George Washington did not chop down a cherry tree! But there is one story which does seem to be true: that of Isaac Newton and the apple.

It is said that Newton was sitting in his garden, one day in the year 1665, when he saw an apple fall from the branch of a tree. (The apple orchard is still there; you can see it for yourself if you go to Newton's old cottage in Lincolnshire.) Why did the apple fall? Newton knew that it was because of the Earth's pull of gravity, and he then realized that the force acting on the apple is the same as the force which keeps the Moon moving round the Earth. It is also the force which keeps the Earth moving round the Sun.

But if the apple falls, why does the Moon stay up? The reason is that the Moon is moving. If you take an object and whirl it round on the end of a string, it will not fall; so long as you keep whirling, the object will go on moving round your hand. The Moon is moving round the Earth at the rate of over 2200 miles (3600 kilometres) per hour, and it keeps on doing so because there is nothing to stop it. It is well above the top of the air, so that there is no friction and nothing to act as a brake. This is why the Moon never comes any closer to the Earth.

In the same way, the Earth is moving round the Sun at over 18 miles (30 kilometres) per second or 66,000 miles (106,000 kilometres) per hour. There is no fear that it will start to fall toward the Sun, because there is nothing to stop it from moving.

Now suppose that you are standing on top of a high tower, and you throw a stone away from you. It will follow the path marked 1 in the diagram, and then fall to the ground. A harder throw means that it will travel further (path 2). If you could throw the stone at a speed of 5 miles (8 kilometres) per second, it would not fall back; it would go right round the Earth – and hit you on the back of the neck (path 3). Of course, nobody can throw a stone at this 'circular velocity', but you can see what will happen if we send up an

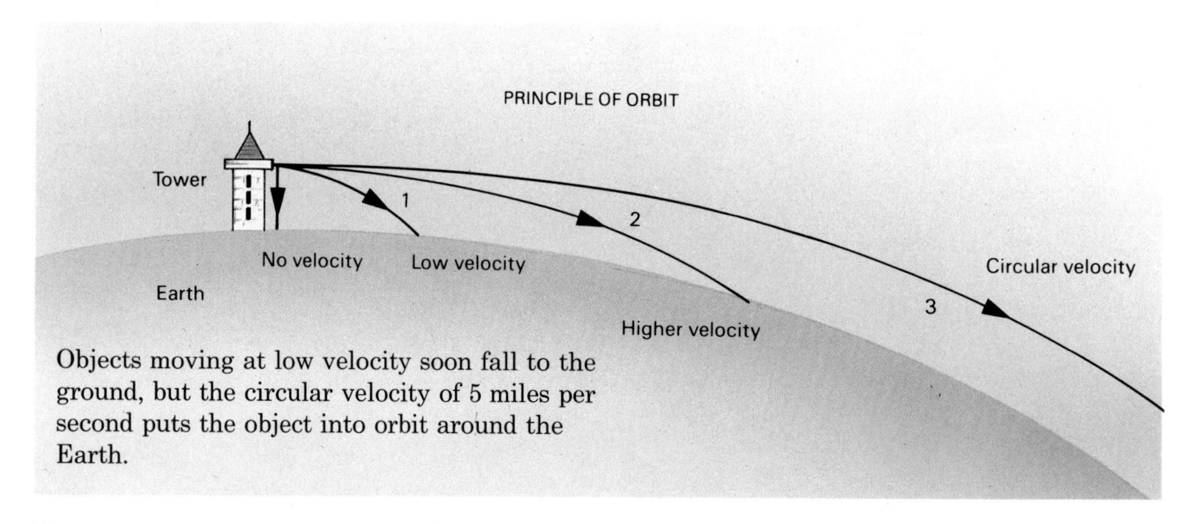

Objects moving at low velocity soon fall to the ground, but the circular velocity of 5 miles per second puts the object into orbit around the Earth.

Sputnik 1, the first artificial satellite to orbit the Earth, was launched by the Soviets in 1957. For almost three months it circled the Earth, over 100 miles up.

object above the top of the air, and then start it moving round the Earth at circular velocity. It will become a man-made moon, or 'artificial satellite'.

Tsiolkovskii knew this, but in his day there was no way in which it could be done. Rockets gave the answer, and on 4 October 1957 the Soviets launched the first artificial satellite. They called it *Sputnik 1*. It went round the Earth at a height of over 100 miles (160 kilometres), spinning round as it did so, making one circuit in less than two hours. It stayed up for almost three months.

Sputnik 1 was about the size of a football. All it carried was a radio set, which sent out the 'bleep! bleep!' signals which were heard by radio operators all over the world. A few weeks later the Soviets sent up a larger satellite, *Sputnik 2*; in early 1958 the Americans had their first success, and since then thousands of satellites have been launched, some of them very large. Because they are lit up by the Sun, they can be seen shining like stars; but unlike stars, they move along slowly. If you see a star-like point crawling across the sky on a dark night, it will be an artificial satellite – unless, of course,it is something much lower down, such as an aircraft.

Sputnik 1 did not stay up for more than a few weeks, because it was not above the very top of the air. There was still some atmosphere around it, and very slowly it was 'braked', until it dropped into the thicker part of the air and was burned away by friction. This also happened with *Sputnik 2* and many of the later satellites, but there have also been many which are so high up that they will never come down. They have become true members of the Solar System.

All the early satellites were sent up either by the Soviets or the Americans, but now other countries have joined in. The satellites have become useful in many ways, and it seems strange now to think that only fifty years ago there were still some people who were quite sure that we would never be able to break free from the pull of the Earth.

A photograph showing the trail of *Skylab* across the sky in 1973. The short bright trail is the planet Mars.

CHAPTER 5
HOW SATELLITES ARE USED

Without the air, we could not live. Life could never have started on the Earth – just as it has never been able to start on the Moon, where there is no air. But it must also be said that there are drawbacks to living at the bottom of an 'ocean of air'. The atmosphere is not only dirty and unsteady; it also blocks out many of the radiations coming from space, just as a sheet of cardboard will block out the beam of a torch.

If you throw a stone into a calm pond, it will set up waves. The distance between one wave crest and the next is known as the 'wavelength'. In the pond, it may be a few inches or centimetres. Light is also a wave motion, but the wavelength is much shorter, so that in everyday life you cannot possibly measure it. The colour of the light depends upon the wavelength; red has the longest wavelength, and then come orange, yellow, green, blue and violet. If the wavelength is longer than that of red light, or shorter than that of violet, we cannot see it, but we can measure these radiations in other ways.

If the wavelength is longer than that of red light, we come to 'infra-red', which you may call heat. Of course, 'heat' and 'light' are not the same thing, but both are radiations, and it is the difference in wavelength which matters. Switch on an electric fire, and you will feel the infra-red, in the form of heat, well before the bars become hot enough to glow. If the wavelength is longer still, we come to radio waves. If the wavelength is shorter than that of violet light, we have ultra-violet (and many people will know the ultra-violet lamps used in hospitals), beyond which come radiations with even shorter

Communication satellites orbit the Earth at a height of 25,000 miles. One journey takes 24 hours, so they seem stationary from the Earth's surface. Three such satellites can give a virtually world-wide TV coverage.

wavelengths which we call X-rays.

Bodies in the sky send out radiations of all kinds, but many of these have wavelengths which cannot pass through the Earth's air; if we want to study them, we must send up our instruments in rockets or artificial satellites.

One of the first important results was the discovery of belts (or zones) of radiation round the Earth. These were found by instruments in the first American satellite, *Explorer 1*, which was launched in 1958. The astronomer who designed the instruments was James Van Allen, so that the zones are now named after him; we have since found that some other planets have zones of the same kind (though the Moon does not). Satellites have also studied the so-called 'cosmic rays', which are not rays at all, but high-speed pieces of atoms rushing in toward the Earth from all directions all the time.

Satellites are also used to take pictures of the Earth from great distances, so that we can see whole cloud systems; the pictures are shown every day on television, and weather forecasting has been much improved.

Communications satellites are also very important. The first satellite of this kind, *Telstar*, was launched in 1965. It was put into a path high above the Earth – about 24,000 miles (39,000 kilometres), or a tenth of the way to the Moon – and it took 24 hours to make one journey round the Earth. This meant that as the Earth spun round, *Telstar* kept pace with it, and seemed to stay in the same position in the sky. A television signal transmitted from America was sent to *Telstar*, and then beamed down to Europe. Today there are many satellites like this.

Tiros 1, the first of the weather satellites that now provide continuous pictures of the atmosphere for weather forecasts.

Telstar itself was very small, and we do not know where it is now. Its power has failed, and of course we cannot see it, but it must still be moving round us, and will keep on doing so until something hits it and destroys it. Some of the other communications satellites have been much larger and brighter, and are used for telephone, radio and television links all over the world.

A few days ago I turned on my television set to watch a cricket match being played between England and Australia. The match was taking place in Sydney, Australia; I was sitting in my study in Sussex, but I could see the game as clearly as if it had been a few yards away from me. Without artificial satellites, anything of this kind would be quite impossible.

CHAPTER 6
THE SPACE TRAVELLERS

The Soviets sent up the first man-made moon; four years later, in 1961, they sent up the first space traveller, Yuri Gagarin. His flight lasted for less than two hours, but he made a full journey round the Earth above the top of the atmosphere.

Before Gagarin's flight there were still some people who believed that manned space flight would never be possible. There were several reasons for this. First, there was the danger of being hit by meteorites – small pieces of rock – which could easily destroy a spacecraft. Then there were the cosmic rays, which, as we have seen, are high-speed particles coming in from all directions; would they be harmful? And there was also the problem of having no 'weight'.

Artist's impression of a meteor shower. Commonly known as shooting stars, meteors are too small to damage spacecraft.

It is true that there are plenty of small bodies moving round the Sun. Most of them are no larger than specks of dust, and we see them only when they dash into the top part of the Earth's air, when they become hot by friction against the atmosphere and burn away to cause the streaks of light which we call shooting stars or meteors. There are also larger bodies, called meteorites when they hit the ground; if you go to Arizona, in the western USA, you can see a large crater which was made by a meteorite many thousands of years ago. Before Gagarin made his trip, nobody was sure whether meteorites would be a real danger, but up to now there have been no hits by any bodies large enough to do real damage. Neither have cosmic rays caused any trouble. This leaves us with the problem of losing one's weight.

Once in space, moving round the Earth in what is called 'free fall', you seem to have no weight at all. Inside his cabin, an astronaut can float about, and there are all sorts of strange effects. For example, you cannot pour out a drink, because the liquid is weightless and will not pour; if you hold out a pencil and let go, the pencil will not drop, but will stay where it is.

It had been thought possible that this lack of weight would make a space traveller sick, just as many people suffer from sea-sickness. Gagarin showed that this was not true. Some of the later spacemen have felt sick or giddy for a while, but not for long; they soon become used to this state of weightlessness or 'zero gravity'.

(*left*) Astronauts Conrad and Kerwin use weightlessness to good effect, as Kerwin examines his partner's mouth.

(*below*) Alan Shepard, America's first man in space, sets foot on the Moon during his second spaceflight, as Commander of *Apollo 14*.

It has often been thought that weightlessness is caused by moving out of the Earth's pull of gravity, but this is not so. The reason is quite different, as can be shown by an easy experiment.

Take a book, and put a quarter on top of it. The quarter is pressing down on the book, so that with regard to the book the quarter is 'heavy'. Now drop the book. While the book and the quarter are falling to the floor, the quarter stops pressing on the book; certainly the quarter is falling, but the book is 'falling away' from underneath it at the same rate, so that with respect to the book the quarter has become weightless. The same would be true if both book and quarter were moving upward at the same time.

Now think about an astronaut inside his spacecraft. He and the spacecraft are moving in the same direction at the same speed; just as the quarter stopped pressing on the book, so the astronaut stops pressing on his spacecraft. He has become 'weightless'. All the astronauts say that zero gravity is not in the least uncomfortable; Gagarin himself once told me that he enjoyed it.

The first American to go into space was Alan Shepard, a few months after Gagarin; since then there have been many more. All astronauts have to go through a very long, hard training. Space travel is dangerous, and several of the astronauts have lost their lives. Gagarin himself was killed in an ordinary aircraft crash some years after his space flight. He had hoped to go to the Moon; it is sad that he was not able to do so.

CHAPTER 7
SPACE STATIONS

After the early space flights, both Russia and America were ready to build larger and more powerful rockets. Before long they were also able to guide their spacecraft, and to bring them together by 'docking'.

If two spaceships are moving in the same path or orbit, they will stay together even though they are moving round the Earth at thousands of miles or kilometres per hour. To see what is meant, think of two ants crawling on the rim of a bicycle wheel. If you spin the wheel, the ants will not fly apart, and one of them can easily crawl up to the other. It is the same with two or more spacecraft; by using their rocket motors, they can dock together, as has often been done. If an astronaut puts on a spacesuit and goes outside his craft, he will be in no danger of drifting away at high speed. The first man to do a 'spacewalk' of this kind was another Russian, Alexei Leonov.

Yet there is always danger. If you start to drift away slowly, you could be in trouble, which is why Leonov was connected to his spaceship by a cord. Since then there have been astronauts who have made 'walks' without safety cords, but all of them have had small rocket motors in their spacesuits, to guide themselves.

I have already said something about Tsiolkovskii, the Russian who wrote about space travel many years ago. He believed that it would be possible to build a space station, put into a path round the Earth at a height great enough to make sure that it was not dragged down into the thicker part of the air. The first true space station, the American *Skylab*, was launched in

(*left*) A possible future space station, with the re-usable shuttle approaching.

(*right*) The Soviet *Soyuz* spacecraft, which carries cosmonauts to the *Mir* space station.

1973, though it was damaged as it went up and its first crew, made up of three astronauts, had to make some repairs to it before they could go inside. *Skylab* was not above the whole of the atmosphere, and it burned away in the upper air some years later; but by then it had been manned by three crews, one after the other, and had shown that space stations could indeed be built.

Skylab was launched by a powerful rocket, and the crew followed afterwards in their own rocket. The next need was for a spacecraft which could be used over and over again. Otherwise, space travel would be very expensive – just as it would be expensive to build a new train for every rail journey.

The first 'reusable' craft was the Space Shuttle, which was able to go into orbit and then return to Earth. It has been said that the Shuttle was launched like a rocket, flew round the world like a spaceship, and landed like a glider. All went well until 1986, when the Space Shuttle *Challenger* blew up a few seconds after launch, and all eight members of its crew were killed.

This was not the first accident in space travel; earlier, four Russians had lost their lives during flights, but it

(*right*) *Skylab*, the first US space station. This was visited by three crews using ferry spacecraft during 1973.

was a tragedy which shocked the world. All Shuttle flights were stopped for some years. Meanwhile, the Russians had been building space stations of their own; the latest of them is *Mir* (the Russian word for 'peace'), and it has been a great success. One spaceman, Yuri Romanenko, has spent 11 months on board *Mir*. At the end of his trip he and his companion landed safely back on Earth, while two other astronauts were sent up to work on *Mir* space station.

We cannot yet be sure whether or not being weightless for long periods will harm the space travellers. Of course they take regular exercise during their flights, but their muscles must be weakened, just as you will find it hard to stand if you have had to spend several weeks in bed, and there could be other problems too. We must wait and see what happens in the future; but at any rate, we cannot doubt that new space stations will be launched before long.

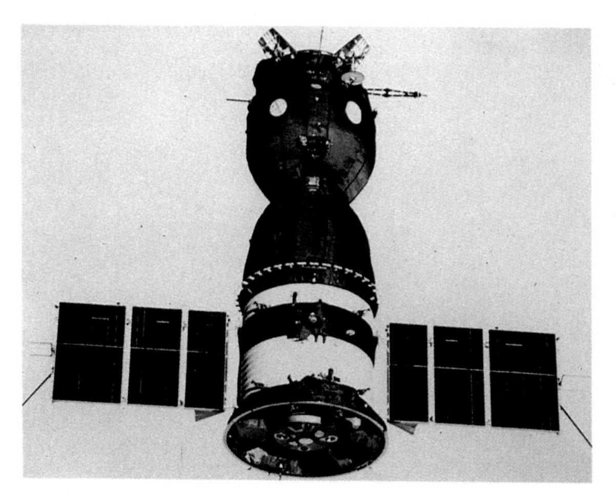

CHAPTER 8

ROCKETS TO THE MOON

If we plan to travel to other worlds, we must begin with the Moon. It is close to us, and it always stays with us as we move round the Sun; a rocket can reach it in only a day or two, whereas it takes many weeks to reach even the nearest of the planets.

When you look at the Moon, it is easy to see that there are dark patches and brighter parts. The dark patches are known as 'seas', but there has never been any water in them; a 'sea' on the Moon is a dry plain. The Moon has no water; neither has it any air, so we cannot hope to find life there.

The Moon has no air because its gravity is so weak. We have seen that the Earth's escape velocity is 7 miles (11 kilometres) per second, but the Moon's

JOURNEY TO THE MOON

1 Launch from Earth

2 Entry into Earth's Orbit

3 Exit from Earth's Orbit

4 Trajectory to Moon

5 Entry into Moon's Orbit

6 Descent to Lunar surface

7 Lunar landing and lift-off

8 Re-entry into Moon's Orbit

9 Exit from Moon's Orbit

10 Trajectory back to Earth

11 Re-entry into Earth's atmosphere

12 Splashdown

(*below*) The *Apollo 11* command and service module in lunar orbit, July 1969.

is only 1.5 miles (2.4 kilometres) per second. Atmosphere is made up of particles which are flying about at high speeds. If an air particle is moving outward at escape velocity, it will break free. This cannot happen with the Earth, but the Moon's pull is so weak that any air it may once have had has drifted away into space. There are no clouds, no storms, no winds and no weather; days on the Moon are always sunny.

On the Moon, a 'day' is much longer than it is at home. The Earth spins round once in 24 hours, so that we have about 12 hours of daylight and 12 hours of night (though this does change according to the time of year; there is more daylight in summer than in winter). The Moon spins round not in 24 hours, but in just over 27 days, so that a 'day' there is almost as long as two Earth weeks, with a 'night' of the same length.

The Moon also takes just over 27 days to make one journey round the Earth. This means that the same side of the Moon is turned toward us all the time. You can show this by walking round a chair, keeping your face turned toward the chair. Anyone sitting on the chair will never see the back of your neck, and from Earth we can never see the 'back' of the Moon.

Before the Space Age, we did not know what the far side of the Moon was like, though we thought that it would be much the same as the side we can see – airless, lifeless, and covered with mountains, valleys and craters. Some of the Moon's craters are over 100 miles (160 kilometres) across, and there are mountains nearly as high as our Everest.

The first Moon rockets, known as Lunas, were sent up by the Russians in 1959. *Luna 1* passed by the Moon without landing there; *Luna 2* crashed in one of the dry 'seas'; *Luna 3*, in October 1959, went round the Moon and sent back the first pictures of the far side. The pictures were taken when *Luna 3* was well beyond the Moon, and were sent back by television. They showed that the far side was indeed much as we had believed, though there were not so many of the dark 'seas'.

The upper stage of *Apollo 17*'s lunar lander returns to lunar orbit from the Moon.

Even after these early rockets, it was still thought that the Moon might be covered with soft dust, in which case it would be dangerous to land there. Luckily, this is not true. First the Russians, then the Americans, sent up spacecraft which landed safely on the Moon without being damaged; they slowed themselves down by using rocket power. There was no danger of burning up during the last part of the journey, because the Moon has no air to set up heat by friction.

Within ten years of the first true space flight, the Americans were ready to send men to the Moon. This was the start of what we will always remember as the Apollo programme.

CHAPTER 9
MEN ON THE MOON

No rocket can yet carry enough fuel to take an astronaut straight from the Earth to the Moon and back again. For the Apollo programme, a different plan had to be worked out.

The first few Apollo flights were 'tests'. It was only with *Apollo 8*, in 1968, that the three crew members actually went round the Moon and had the first direct views of the far side, which we can never see from Earth. *Apollo 9* was another test; *Apollo 10* took three more astronauts round the Moon – and all was ready for the first landing, with *Apollo 11*. The three astronauts were Neil Armstrong, Edwin Aldrin and Michael Collins. Armstrong and Aldrin were to land on the Moon, while Collins stayed in the other part of the spacecraft.

The launching, in July 1969, was carried out by a very powerful rocket, which soon used up all its fuel and broke away, falling back to the ground. The second rocket, which it had taken up, then went on with the journey, and it was this which put the true spacecraft on its way to the Moon.

When they were near the Moon, Armstrong and Aldrin went into the *Eagle*, or lunar module, a separate, smaller spacecraft which they had brought with them, and which was to be used for the landing itself. Leaving Collins behind to go on circling the Moon, Armstrong and Aldrin fired the motor of the *Eagle*, and dropped down toward the Moon. They slowed themselves down by firing their rocket motor to act as a brake, and then Armstrong brought the *Eagle* down gently on to the grey plain known as the Sea of Tranquillity. Millions of people watching on television or listening on radio all over the world heard him say: 'The *Eagle* has landed.' A few hours later he went down the ladder on to the surface of the Moon, followed by Aldrin a few minutes later.

They found a strange world. The sky was black, because there was no air to scatter the Sun's light around and make the sky blue; there were pits and small hills everywhere, and they could see the Earth, looking much larger than the Moon does to us. Because of the weak gravity, they had only one-sixth of their usual weight, so that they looked as though they were moving around in slow motion. They set up their instruments, and then went back into the *Eagle* to prepare for their return.

The *Eagle* had only one motor; if this motor had not worked properly, the astronauts would have been left on the

The lunar rover or 'Moon Car' of *Apollo 17*, which enabled the astronauts to explore much further from the lander than was possible on previous missions.

Moon with no chance of being rescued. Luckily, all was well. The motor fired; Armstrong and Aldrin were lifted away from the surface – leaving behind the bottom part of the lunar module, which they had used as a launch pad – and docked with Collins in the orbiting part of the *Apollo*. The return journey was made in the main spacecraft, and the final landing in the sea marked the end of man's first journey to the Moon.

(*left*) *Apollo 11* astronaut Edwin 'Buzz' Aldrin photographed during the first Moon walk.

(*below*) A map of the three journeys made by the *Apollo 17* astronauts on the Moon's surface.

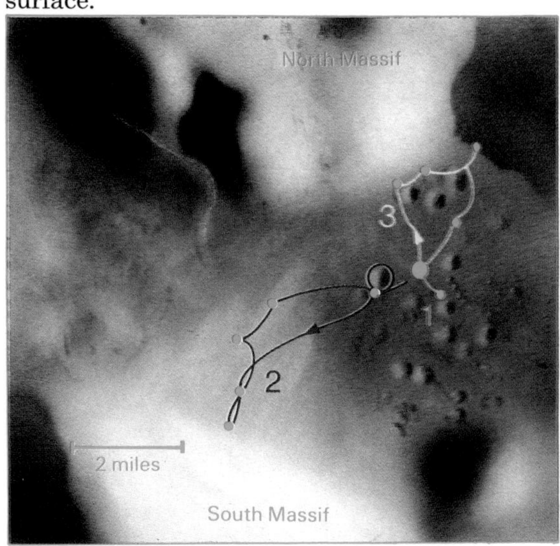

Apollo 12 followed in December of the same year. This time the astronauts, Charles Conrad and Alan Bean, came down close to an old American spacecraft, *Surveyor*, which had been sent to the Moon some years before. They were able to break pieces off the *Surveyor* and bring them home to be studied.

Apollo 13 was nearly a disaster. On the outward trip there was a fire in the engine section of the spacecraft, and it was not possible to land on the Moon as had been planned. The astronauts were able to go round the Moon and return safely, but it was a very narrow escape.

There were four more Apollo missions. On the last three the astronauts drove around the Moon, using special electrically-powered 'cars' which they had brought with them. The 'cars' had to be left behind; they are still on the Moon, and one day, no doubt, a new astronaut will go up to them and drive them away!

The last Apollo was No. *17*, in December 1972. Since then no men have been to the Moon, but we may hope that there will be more journeys there before the year 2000.

CHAPTER 10
LIVING ON THE MOON

We know much more about the Moon today than we did before the Apollo flights. We have found out that the surface is solid enough to bear the weight of a spaceship, and we have been able to study lunar rocks – not only those brought back by the astronauts, but also some which were brought back by unmanned Russian rockets which landed on the Moon, collected rocks and then came back to Earth. The instruments left behind by the astronauts went on working for several years, and told us a great deal. On the Moon, where there is no air, we can study all the radiations coming from space.

From the Moon, the Earth looks large and bright. Because the sky is always black, stars can be seen in the daytime if the astronaut shields his eyes from the glare of the Sun. Nothing moves, because there is no wind; the flags set up by the Apollo astronauts

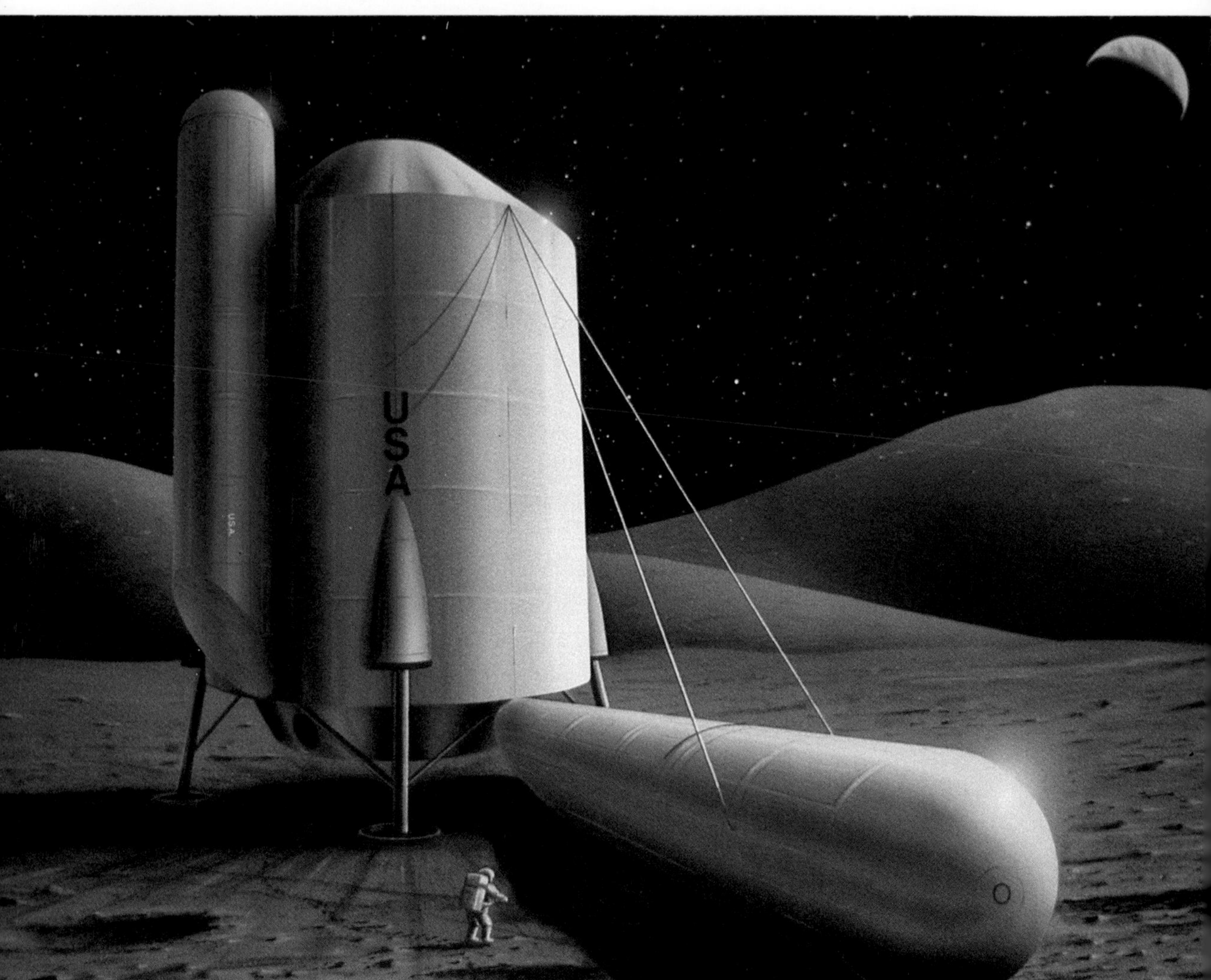

Artist's impression of a Lunar Base, which could be in operation on the Moon within the next decade or so.

do not wave about, and, of course, there are no clouds.

Though the US program was held up by the disaster with the *Challenger* shuttle, and the Russians have not yet tried to send men to the Moon, plans for a Lunar Base are already being made. However, one thing which we can never do is to walk about on the Moon without wearing a very special spacesuit. Astronauts must be able to breathe, and so they have to carry supplies of oxygen gas with them. But their suits must also be tough, because there is no air pressing down – and if the suits were not strong enough, the astronauts would die at once,

This means that anyone going to the Moon must always wear a suit, or else stay inside a spacecraft or a lunar base. One idea for a base is to make plastic domes, kept blown up by the air inside them like large balloons. This may well be the shape of a future base, though the first bases will be much more simple.

There is no reason to doubt that a Lunar Base could be set up within the next ten years or so. It will be used for scientific work; on the Moon, it is possible to carry out many experiments which could not be tried on the Earth. We hope that we will learn more about the way in which the human body works; there will be laboratories for chemistry and physics, and, of course, observatories for use by astronomers. On the Moon's far side, from which the Earth can never be seen, there will be radio observatories. No radio waves from Earth will be received there, because radio waves cannot travel through the solid body of the Moon, and there will be nothing to interfere with the work of the scientists.

Apollo 12 astronaut Al Bean removes part of the *Surveyor 3* TV camera.

The first astronauts to build the Lunar Base will be carefully chosen, but in the future ordinary people should be able to go there, even if the idea of taking holidays on the Moon is a long way away yet.

We have made great progress since the start of the Space Age. Indeed, we have learned more quickly than anyone could have hoped; Alan Shepard, who made his first short flight in 1958, was one of the *Apollo 14* astronauts who went to the Moon. Yet the Moon is only a start. We had to go there first because it is so near, but now we can look further into space – to the other worlds of the Solar System.

CHAPTER 11
TRAVEL TO THE PLANETS

As soon as we look beyond the Moon, we find that things become much more difficult. The worst problem is that of the distance we have to travel. The planets go round the Sun, not round the Earth, so that they do not stay close to us in the way that the Moon does; even the closest planet – which is Venus, not Mars – is always at least a hundred times as far away as the Moon. If it takes a rocket, say, three days to go from the Earth to the Moon, it would take almost ten months to reach Venus even if it could go by the shortest path.

But a rocket cannot go by the shortest path because, as you will see, it would need more fuel than it can carry. This means that space travellers must be ready for a very long journey.

The Sun's family, or Solar System, is made up of two main parts. First we have four small, solid planets – Mercury, Venus, the Earth and Mars; the Earth comes third in order of distance from the Sun. Outside the path of Mars we have a wide gap in which move thousands of very small worlds called minor planets or asteroids. Beyond these we find the four very large planets Jupiter, Saturn, Uranus and Neptune, together with one small body, Pluto.

Artist's impression of Venus. The thick cloud layer prevents us seeing the planet's surface.

First let us look at the two planets closest to the Earth: Venus and Mars. Venus is closer to the Sun than we are, and is about the same size as the Earth. It has a thick, cloudy atmosphere, and in our sky it shines more brightly than any other planet or star. It is 67 million miles (108 million kilometres) from the Sun, and it takes almost 225 days to go once round the Sun, so that this is also the length of Venus' 'year'.

Because the Sun can shine on only half of a planet at one time (just as the beam of a torch in a dark room can light up only half a football), Venus seems to show changes of shape like those of the Moon. When Venus is

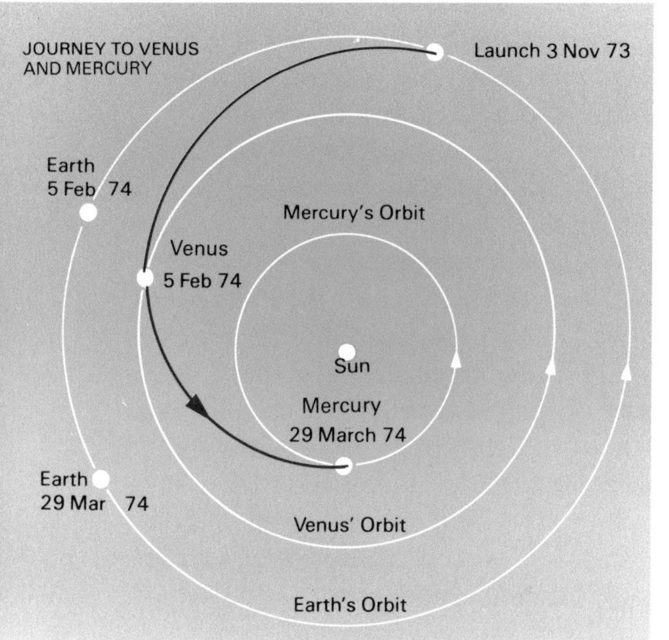

The trajectory of the *Mariner 10* mission to Venus and Mercury, launched in 1973.

between the Earth and the Sun it is 'new', and we cannot see it, because its dark side is turned toward us – unless it passes exactly in front of the Sun, and appears as a dark spot against the Sun's face, which will not happen again until the year 2004. When closest to us, Venus is about 24 million miles (38 million kilometres) away. It would be useful to send a rocket straight across the gap – but, as we have seen, this would need too much fuel.

What we have to do is launch our spaceship by using a rocket, in the usual way, and then 'slow it down' with respect to the Earth. We fire the rocket motors in the opposite direction to that in which the rocket itself is moving. This means that the motors will slow the rocket down instead of speeding it up. The spacecraft will start to swing in toward the Sun, and it can be made to reach the path of Venus at a meeting point which has been worked out. This was first done by the United States with the spacecraft *Mariner 2*, which was launched in August 1962 and passed by Venus in the following December. Since then there have been several others, both American and Russian, some of which have even made gentle landings on Venus.

Next let us turn to Mars, which is further away from the Sun than we are. It never comes as close to us as Venus; it is always at least 34 million miles (58 million kilometres) away. It is, on average, 141 million miles (228 million kilometres) from the Sun, and it has a 'year' equal to 687 Earth days.

To reach Mars, we must take up our spaceship and then give it extra speed. It will then move faster than the Earth, just as an object being whirled round on the end of a string will swing outward if you loosen your grip and make the string longer. The spacecraft can then be made to reach the path of Mars some months later. The first time this was done was by America's *Mariner 4*, which was sent up in November 1964 and bypassed Mars in July 1965.

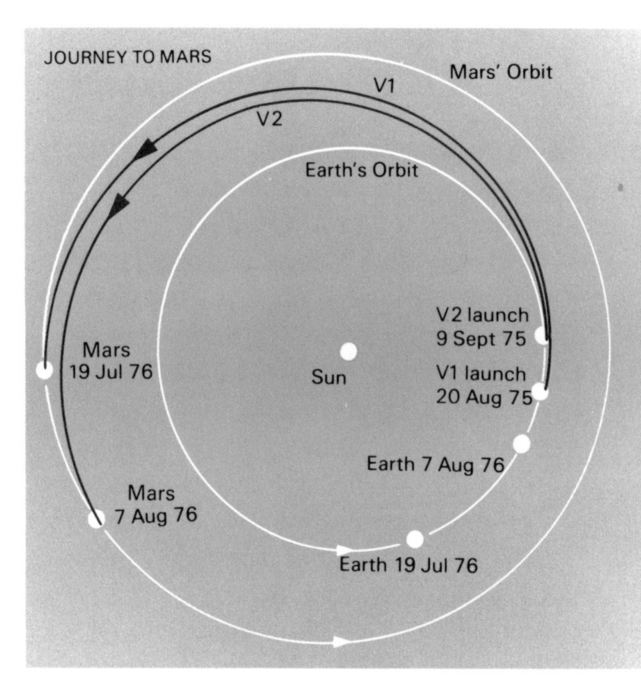

The trajectories of the two *Viking* missions to Mars launched in 1975.

Much less fuel is used by this method. Once the spacecraft has been put into its path, there is no need to give it any more power from its motors. It will simply move on until it comes close to its target planet. When it arrives, it can either be left to go on moving in a path round the Sun, or else given a burst of power from its own motors to put it into a path round the planet. One day, we may be able to make rocket motors which can carry enough fuel to go by the shortest route; until then, we will have to put up with long journeys.

CHAPTER 12
ROCKETS TO VENUS

Because Venus is closer to us than any other planet, there were good reasons for sending spaceships there. Another reason was that Venus and the Earth are about the same size, so that before the Space Age it was thought that there might be life. Yet we did not know much about Venus, because the atmosphere is always cloudy, and no telescope can show us the true surface.

Some astronomers believed that Venus had large seas, with not much land; others thought that there might be no water at all. It seemed clear that Venus must be hot, partly because it is close to the Sun and partly because we knew that the atmosphere was not the same as ours. Our air is made up mainly of two gases called oxygen and nitrogen; that of Venus contains a great deal of a heavy gas known as carbon dioxide – the gas which makes cola drinks fizzy. It was even thought that Venus might have seas of fizzy water!

The first successful spacecraft to Venus, *Mariner 2*, showed that there can be no seas, because Venus is so hot that water would quickly boil away. Next, the Russians sent up spacecraft which were planned to land on Venus and go on sending back pictures after arrival. The first few of these 'landers' failed; we know now that they were squashed by the thick atmosphere before they reached the surface. At last, in December 1970, the Russian *Venera 7* spacecraft landed safely, and sent back signals; but it was not until the landing of another spacecraft, *Venera 9*, in October 1975 that we received our first picture direct from Venus. Since then other spacecraft have landed there, but none of them can go on working for long, because Venus is too hot and the climate is too unfriendly.

Venus is a strange place. The surface is covered with rocks which are so hot that they glow a dull orange-red. The clouds are not at all like our rain clouds, and contain drops of an unpleasant acid known to chemists as sulphuric acid. (Many school science rooms have bottles of sulphuric acid, but never touch it, because it will burn your skin.) The thick atmosphere would press down on you so heavily that you would need a very strong spacesuit to avoid being crushed. You could never see the Sun or the Earth through the clouds – and the day is much longer than ours; when the Sun rises, it will not set again for a period equal to over eight Earth weeks.

Spacecraft moving round Venus have also been able to tell us that there are tall mountains, deep valleys, and two large highland patches. There also seem to be active volcanoes, and there may be thunder and lightning all the time. We cannot believe that any sort of life can exist there.

The surface of Venus, photographed by the Russian soft-lander *Venera 13.*

Mariner 10, which flew past Venus and Mercury sending back information and photographs.

It is possible that thousands of millions of years ago, when the Sun was less hot than it is now, there may have been seas on Venus; but as the Sun grew hotter, the water boiled away and the planet turned into a dust desert. There seems no chance that astronauts will be able to land, but we should be able to send down unmanned spaceships and collect rocks from the surface. When this is done, we will be able to find out whether there are any signs of past life.

In 1974 one of the American spacecraft, *Mariner 10*, sent back pictures of Venus and then went on to pass by the other inner planet, Mercury, which has almost no air, and has craters and mountains very like those of the Moon. Mercury, only 36 million miles (58 million kilometres) from the Sun, is very hot; if you could put a tin kettle on the rocks during the daytime, the kettle would melt. No spaceships have been sent there since the flight of *Mariner 10*, but we may be sure that Mercury, like Venus, is a world without life.

Artist's impression of a landscape on Venus, showing active volcanoes. The rocky surface is so hot that it glows orange and red.

CHAPTER 13
MARS AND ITS MOONS

When Mars is at its closest to us, it looks like a brilliant red star. With a telescope, you can see that the surface is red with dark patches here and there, while the planet's poles are often covered with white 'caps'. The red parts of Mars are usually called deserts, though they are not like our deserts such as the Sahara. They are cold, not hot, and you certainly will not find camels and palm trees. It was once thought that the dark patches were seas, but we now know that this is wrong. There are no seas on Mars, and the dark patches are simply places where the red dust has been blown away by the winds, so that we can see the dark rocks underneath. The white polar caps, shown in the picture, are made up of ice.

Mars is more like the Earth than any other planet, and it was once thought that there might be life there. Some astronomers used large telescopes to look at Mars, and then made drawings of it. They drew thin lines upon Mars which they called 'canals', and believed that they had been built by the inhabitants to carry water from the polar ice through to the warmer parts of the planet. It was only when the first spaceships sent back pictures from Mars that we were sure that there were no canals.

Mars has high mountains, large craters, deep valleys and giant volcanoes. One of these volcanoes, known to us as Olympus Mons (in English, Mount Olympus) is three times as high as our Everest, and at its top is a crater big enough to hold the whole of a city such as Vancouver or Boston.

In May 1971 the Americans launched a new spacecraft, *Mariner 9*. It reached Mars in the following November, and was put into a closed path round the planet, so that it could take photographs of the whole of the surface. Yet when it first arrived, it

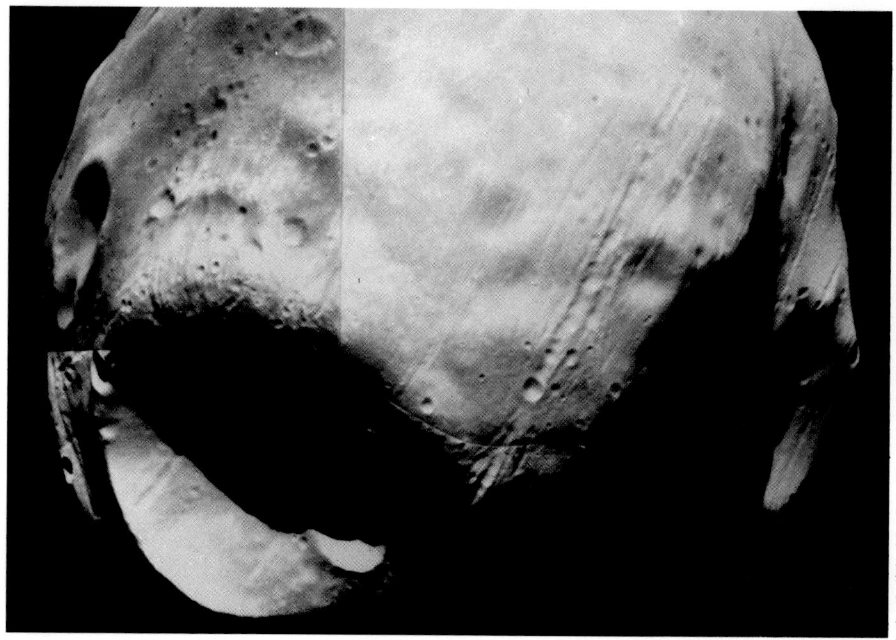

A Viking photograph of Phobos, the larger of Mars' two moons. Both Phobos and Deimos (the second moon) orbit close to the planet and are less than 20 m (30 km) across; they would not give much light during the nights. They are made of dark rock, with craters and hills.

Artist's impression of Mars from nearby, showing both the polar caps.

One of the first photographs of the Martian surface received from *Viking 1*.

could see very little, because there was a great dust storm going on, and all that could be made out was the top of a dusty layer. It was some weeks before the dust cleared away, and *Mariner 9* could begin its real work. It went on sending back pictures for almost a year.

The two Viking spaceships of 1975 were even more important. They took seven months to reach Mars, and then began to move round the planet in the same way that *Mariner 9* had done. Each Viking was made up of two parts, one of which could break away at a command from Earth and drop down on to the surface, braked partly by rocket power and partly by parachute. When the landers touched down, they were able to send back pictures as well as making measurements. *Viking 1* landed in a red plain known as Chryse; *Viking 2* came down 4610 miles (7420 kilometres) to the north east in another red plain, known as Utopia.

The Vikings told us that Mars is always cold. The atmosphere is very thin, and is made up of carbon dioxide; though the winds can blow at speeds of several hundreds of kilometres per hour, they have very little force, and would not be able to knock you over. There are thin clouds, some of them made up of ice, and there is plenty of dust in the atmosphere. This makes the daytime sky pink instead of black, as on the Moon. The 'day' there is about half an hour longer than ours.

Both the Viking landers carried out searches for life. They were able to send out scoops to collect rocks and dust from the surface; this material was brought into the spacecraft, and carefully studied. No signs of life were found, which was disappointing. We cannot yet be sure that Mars is completely lifeless, but it does seem probable.

Though we could not live on Mars without wearing spacesuits or staying inside our rockets or bases, it should be possible to go there, and plans for the first trips are already being made.

CHAPTER 14
THE MARTIAN BASE

Many stories have been written about travel to Mars. Even though there are no Martians, we know that we are not going to find a scorching-hot desert or a dense, poisonous atmosphere. We are also fairly sure that there will be plenty of ice, not only at the poles. The Mariner and Viking pictures show old riverbeds, so that there must have been running water on Mars a long time ago, and we have every reason to think that there will be ice below the surface rocks. If so, we can melt this ice to give us all the water we want.

The main trouble, of course, is that the atmosphere is so thin. Because Mars is smaller and less heavy than the Earth, it has a lower escape velocity – only 3 miles (5 kilometres) per second – so that it has lost most of the air it once had, and astronauts will never be able to walk about in the open without their spacesuits. Moving will be easy; on Mars, you will have one-third of your Earth weight.

Dried-up riverbeds on the Martian surface, photographed by *Mariner 9*.

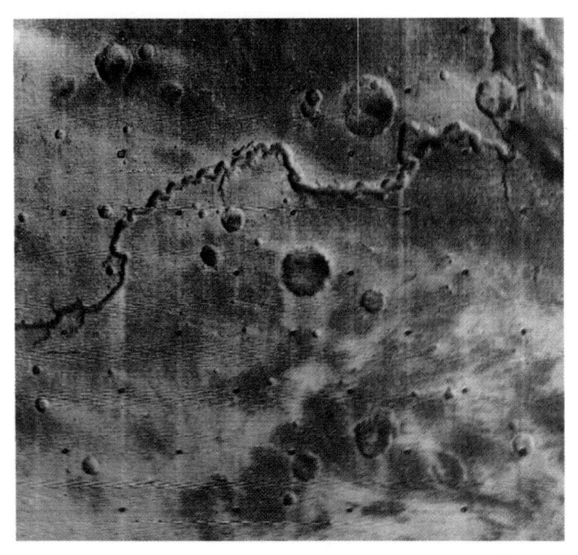

Because a journey to Mars will take months, we cannot make a quick 'there and back' trip, as we can with the Moon. The astronauts will have to spend several months on the planet before Mars and the Earth are in the right places for a return journey, so that the whole trip will last for over two years. The astronauts may go first to one of the tiny moons. Phobos or Deimos, though they may decide to go straight down on to Mars itself.

Landing should not be a real problem, and the surface is certainly quite strong enough to bear the weight of a spaceship. Building a Base will be a different matter. One early plan was to have large plastic domes, kept blown up by the air inside them, in which the astronauts will be able to live and work without having to wear spacesuits. They will use 'air-locks' to go in and out, because it will be very important not to let the air leak away.

The astronauts will have to bring air, food and water with them, but they will then have to make new supplies from what they can find on the planet. Growing plants in the open will not be possible, but the domes could be used in the same way as we on the Earth use greenhouses; oxygen could be made by treating the rocks, and, as we have seen, there is probably plenty of ice to provide water.

There would be endless work to do. Mars is a perfect place for a scientific workshop, including an observatory; it will also be of great value to find out how living things can manage under the weak gravity, and doctors will learn a great deal which will help to save lives on Earth.

Outside the Base, the cold will be

A Viking photo of sunset on Mars. The Sun goes down more slowly than on Earth.

bitter except near the middle of the day; the temperature drops far below freezing point in the afternoon, and the nights even at the equator are much colder than anywhere on the Earth. Land travel should not be difficult, but flying will be more of a problem, because aircraft will not be able to work easily in the thin atmosphere.

Inside the Base, things should be much more comfortable. No doubt there will even be games rooms – though it will seem strange to play football or tennis when nothing has more than one-third of its Earth weight, and during a cricket match a batsman could hit a ball high enough to hit the roof of the dome.

During a flight to Mars an astronaut will, of course, be weightless, and even when on the planet he will still feel very light. When he comes home to the Earth, will his muscles have become so weak that he will be unable to stand up? We do not yet know. Neither do we know whether a boy or girl born on Mars will ever be able to get used to the stronger pull of the Earth. These are problems for the future; but if they can be solved, there seems no reason why we should not set up a Martian Base within the next twenty or thirty years.

Artist's impression of an early landing on Mars – the first step towards a full-scale Martian Base.

CHAPTER 15

INTO DEEP SPACE

Beyond Mars we come to the asteroids or minor planets. There may be at least 40,000 of them, but only one – Vesta – is ever bright enough to be seen without a telescope, and even Vesta looks like nothing more than a faint star.

When plans were being made to send spaceships to the outer planets, there were worries about what might happen in the asteroid belt; if a spacecraft were hit by a piece of rock as large as an orange, it would be destroyed. Even though the rock might be so small, it would have a great deal of force because of its quick movement – just as a marble will hurt you if you throw it hard against your arm. Luckily this has not happened with any of the four deep-space probes so far launched, and it is now believed that the danger is slight. All the asteroids are small, and most of them are only a few kilometres across. None can hold on to any air.

Jupiter, the largest of all the planets, moves round the Sun at a distance of 483 million miles (778 million kilometres), taking almost 12 years to make one full journey. It is big enough to swallow up over a thousand Earths, and it has a strong pull. It is very bright, and can be well seen for several months in every year. Next in order is Saturn, the planet with the rings; it is smaller than Jupiter, but much larger than the Earth, so that it looks like a fairly bright star. It is 886 million miles (1430 million kilometres) from the Sun, and its 'year' is over 29 times as long as ours.

The other three outer planets are much fainter. Uranus and Neptune are each about 30,000 miles (50,000 kilometres) across, but Pluto is smaller than our Moon, and is so dim that powerful telescopes are needed to show it at all.

Sending a rocket even to Jupiter, the nearest of the giant planets, takes a long time. The first success came with the NASA spacecraft *Pioneer 10*, which was launched in March 1972 and passed by Jupiter in December 1973, so that its journey took the best part of two years. *Pioneer 10* sent back good pictures of Jupiter, and so did its twin, *Pioneer 11*, which followed a year later.

When *Pioneer 11* had been past Jupiter, it moved on to pass by Saturn,

Artist's impression of *Pioneer 10* passing an unknown planet. The Pioneer missions may show there are planets in deep space.

Artist's impression of a Voyager spacecraft passing Saturn and Dione, one of its moons, on its way to Uranus and Neptune.

which is much further away. It did not reach Saturn until 1979, over six years after it had been launched. Neither of the Pioneers went close to any other planet, and both are now on their way out of the Solar System. They will never come back, because they are moving too fast. This means that they will simply go on moving away until they are hit and broken up by some solid body, which may not happen for thousands or even millions of years. At present we are still picking up their radio signals, but we are bound to lose touch with them before the year 2010.

Guiding a deep-space probe is not at all easy. If its path is to be altered, the only way to do so is to send out a radio command from the Earth. By the time a spaceship has moved out as far as Saturn, a radio signal will take several hours to arrive even though it is travelling at the same speed as light – 186,000 miles (300,000 kilometres) per second. Also, spacecraft to Mars or Venus can use power from the Sun to make their instruments work, but this cannot be done beyond Jupiter, because there is not enough sunlight. A small atomic power plant has to be used instead. The strength of the signal sent back is very small – not nearly enough to light up the bulb of a flash light for a single second – but it can be used to transmit pictures as well as measurements of all kinds.

Both the Pioneers were successful. Next came the two Voyagers, which are the most remarkable spacecraft ever made up to the present time.

CHAPTER 16

THE VOYAGERS

Because most of the journey from the Earth to another planet has to be done in free fall, without using power, it is bound to take a long time. Luckily things were made easier at the time of the Pioneer and Voyager launches because the four giant planets were spread out in a curve, as shown in the diagram. This made it possible to send a spacecraft from one planet to another – a method which some people have called 'interplanetary snooker'!

Voyager 1 was launched in September 1977, and went to Jupiter, making its pass in March 1979. As the spacecraft approached Jupiter it was pulled onward by Jupiter's strong gravity, and was speeded up. This meant that it could be sent on toward Saturn. It was November 1980 by the time that Saturn was reached, so that the journey there had taken a little over three years; without Jupiter's help, the trip would have taken much longer. After leaving Saturn, *Voyager 1* went on moving away from the Sun. Like the Pioneers, it is travelling so fast that it will never come back.

Voyager 2 was launched a few days before *Voyager 1*, but it did not reach Jupiter until July 1979. It then used Jupiter's pull to send it on to bypass Saturn in August 1981, but this was not the end of its mission. Saturn's gravity was used to send it on to the next planet, Uranus, in January 1986; then the pull of Uranus speeded it up again, so that it will reach the last of the giant planets, Neptune, in August 1989, before leaving the Solar System for ever.

Jupiter is not solid and rocky. Its surface is made up of gas, so that there is no hope of landing there. Though it has a very long 'year', it spins round in less than ten hours. Through a telescope it shows cloud belts and white spots; there is one famous feature known as the Great Red Spot, because of its colour. The Voyagers showed that the Red Spot is a whirling storm, and that the winds on Jupiter are very strong. Though the outer clouds are cold, the planet is hot inside, and it is surrounded by belts of deadly radiation which would quickly kill any astronaut foolish enough to go close in. There are four large moons as well as 12 small ones. One of the large moons, Io, has a red surface, with active volcanoes; of the others, Ganymede and Callisto are ice-covered and cratered, while Europa has an icy surface which is as smooth as a snooker ball.

Saturn is the most beautiful of the planets, because of its rings. The rings are made up of small pieces of ice, spinning round the planet like tiny

A *Voyager 1* photograph of Saturn, which it passed in November 1980.

A model of the Voyager spacecraft, at the Jet Propulsion Laboratory.

moons. Saturn's surface, like that of Jupiter, is made up of gas.

Saturn has no less than 17 known satellites, and one of these, Titan, is of special interest because it has a thick, cloudy atmosphere. On the surface there may be seas, though they will be seas of what is called 'methane' rather than water. Titan is unlike any other world in the Solar System, but we do not expect to find life there, because Titan is too cold.

The two outer giants, Uranus and Neptune, also have surfaces made up of gas; Uranus is green, Neptune blue. Uranus has ten thin, dark rings, quite unlike the bright, icy rings of Saturn. *Voyager 2* showed them well, and also sent back pictures of the larger satellites. One of these satellites, Miranda, has tall ice cliffs together with mountains, craters, and even a large area which looks rather like a racetrack!

It is a pity that neither of the Voyagers went close to Pluto, which was discovered only in 1930. Pluto is very small, and seems to be made up of a mixture of rock and ice. As yet we do not know much about Pluto or its satellite, Charon, but it is certainly very cold, with almost no atmosphere. From Pluto, the Sun would look like nothing more than a very brilliant star.

The Voyagers have done all that had been hoped of them. They are still working well, even though they have been in space ever since 1977, and we hope to keep in touch with them for some years yet.

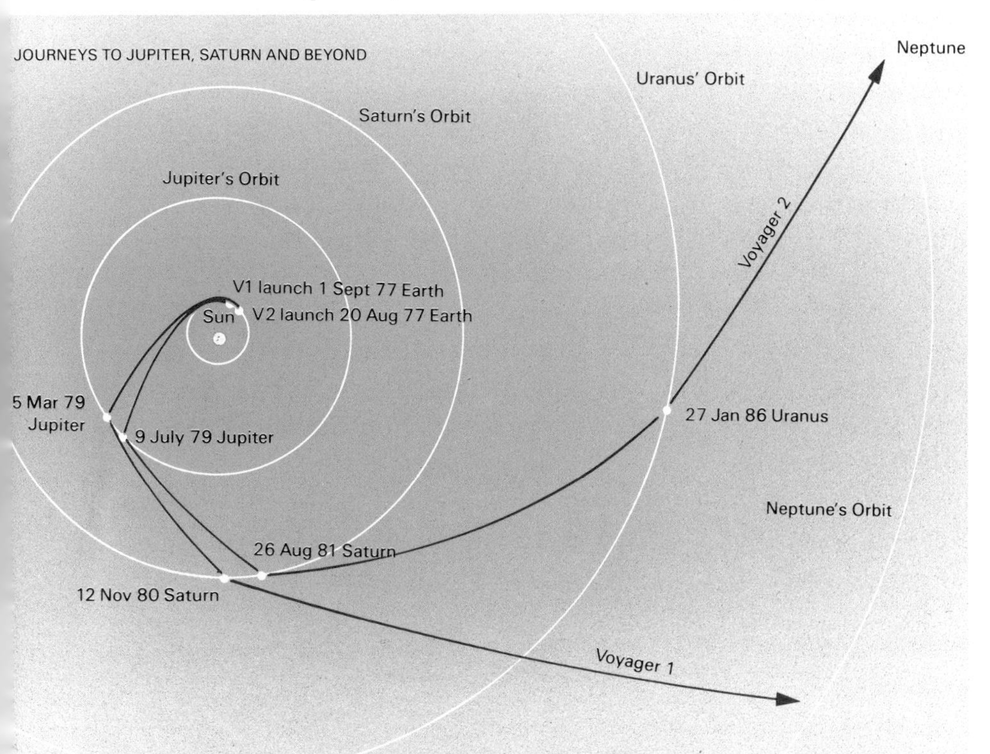

The two Voyager spacecraft were sped on their way by the gravity of each planet they passed. *Voyager 2* – which set off first – was able to visit four planets in all.

CHAPTER 17
INSIDE A COMET

As we have seen, there are other bodies in the Solar System beside the Sun, the planets and their satellites. On a dark, clear night you can often see shooting-stars or meteors, which are no larger than specks of dust, and which burn away in the Earth's upper air. We now know that meteors come from comets.

A comet is not nearly so important as it may sometimes look. Its only solid part is its centre, known as the nucleus, which is made up of ice, and is only a few miles or kilometres across. A comet moves round the Sun, but its path is not almost circular, like those of the planets; it travels in a long, narrow orbit – and because it shines only by reflecting the light of the Sun, we can see it only when it is moving in the inner part of the Solar System. Note, too, that it is millions of miles beyond the Earth's air, so that you cannot see it moving. If you watch something moving quickly across the sky, it cannot be a comet.

Really bright comets are seen now and then, but most of these take hundreds or even thousands of years to make one journey round the Sun, so that we never know when to expect them. The only bright comet which is seen regularly is known as Halley's Comet, because it was carefully studied, 300 years ago, by the great English astronomer Edmond Halley. It comes back every 76 years, so that it was seen in 1835, 1910 and again in 1986 – though of course it stayed in view for some years to either side of those dates. It will be back again in 2061.

In 1986 Halley's Comet could be seen without a telescope for some months, but it never became bright, as it had done in 1910, because it was on the far side of the Sun when it ought to have been at its best. It had a 'head', made up of gas which had been sent out by the nucleus, and a 'tail', made up of thin gas and very small particles streaming away from the head. Both the head and the tail developed when the comet came closest to the Sun, and the ice in its nucleus began to boil off.

Artist's impression of a comet nucleus – an ice block trailing gas and particles.

Plans had been made to send rockets to Halley's Comet, and there were five spacecraft altogether, two Soviet, two Japanese, and one European. The Soviet and Japanese spacecraft passed by the comet, but the European probe, named *Giotto* after an artist who had painted the comet hundreds of years ago, went right inside the head, and was able to take close-up pictures of the icy nucleus. The real surprise was that instead of being bright, the nucleus was covered with a dark layer.

Because a comet has so much 'dust', there were fears that *Giotto* might be hit and put out of action. Just before it passed closest to the nucleus it was indeed hit by a piece of material no larger than a grain of rice, and its signals were cut off for a while; the camera never worked again, but *Giotto* came out of the comet safely, and is still sending back radio signals. It will be back near the Earth in 1990.

Comets lose some of their gas and dust every time they come close to the Sun, so that they do not last for as long as planets; we have seen some comets break up, and others have simply faded away. But a really brilliant comet is a wonderful sight, and it is a pity that so few of them have been seen in the past few years.

Artist's impression of Halley's Comet seen from Australia in April 1986.

Artist's impression of spacecraft *Giotto* encountering Halley's Comet in 1986.

It is thought that a very long way from the Sun – far beyond any of the planets – there is a whole 'cloud' of comets, known as Oort's Cloud after the Dutch astronomer who first suggested it. If one of the comets in the cloud is pulled out of its path, it will fall inward toward the Sun, and we will be able to see it. It may then go back to the cloud, or it may be caught by the pull of a planet (usually Jupiter) and put into a smaller orbit. This is what happened to Halley's Comet.

If Oort's Cloud really exists, as most astronomers believe, it marks the edge of the Solar System. The stars are so much further away that we cannot hope to send rockets to them; for the moment, at least, we must be content to explore our own part of the universe.

CHAPTER 18

OTHER SUNS – AND OTHER EARTHS?

As we have seen, the Sun is a star. There are many other stars like it, but to us they look like tiny points of light, because they are so far away.

To see what is meant, let us build a scale model. Suppose that we make the distance between the Earth and the Sun one inch or 2.5 centimetres, as shown by this line ————— . On the same scale, how far away must we put the nearest star beyond the Sun? The answer is – over 4 miles (6 kilometres). Most of the other stars are much more distant still.

To give these distances in miles or kilometres would be clumsy, just as it would be clumsy to give the distance between, say, New York and Washington in inches or centimetres. (Work out the answer, and you will see what I mean!) Instead, astronomers use a unit called a light year. Light moves at the speed of 186,000 miles (300,000 kilometres) per second, so that in a year it can cover almost 6 million million miles (9.6 million million kilometres). This is the light year, which is a measurement of distance, not of time. Light can reach us from the Sun in only 8.6 minutes, but the nearest star beyond the Sun is over four light years away.

Because the stars are suns, there is no reason why they should not have families of planets, so let us think about the way in which stars are born and how they die.

It is thought that a star like the Sun begins by being formed out of a 'cloud' of dust and gas in space. We can see many of these clouds, known as 'nebulae'; there is one in the star group of Orion, which you can see without a telescope even though it is more than a thousand light years away from us. As the new star becomes smaller, it also becomes hotter, and at last it begins to shine. It goes on shining for a very long time, as our Sun is doing now, but it cannot last for ever. It will run out of energy and will swell out, becoming cool and red. After this, it will shrink down into a very small, very heavy body, and will shine feebly for many more thousands of millions of years before it becomes cold and dead.

The Sun's planets, including our Earth, were formed out of a cloud of dust and gas which used to surround

the Sun, around 4500 million years ago. Other stars of the same kind as the Sun have probably formed planets of their own. We cannot be sure, but one of the artificial satellites has given us a clue. The satellite was called IRAS – the Infra-red Astronomical Satellite – and it was launched in 1983, after which it went on working for almost a year before its radio transmitters failed. IRAS carried instruments to look for very cool gas and dust which would not be hot enough to shine, but would radiate in infra-red. Cool material was

Artist's impression of a possible planetary system detected by IRAS around the star Beta Pictoris.

(*left*) The Infra-Red Astronomy Satellite (IRAS)

found around some stars, one of which was Vega, which as seen from the USA is almost overhead during summer.

It is quite likely that these 'cool clouds' around stars will form into planets, if indeed they have not already done so. In this case, there is no reason why some planets of other stars should not be very like our own Earth.

Unfortunately, we cannot see them. Remember that a planet is much smaller than a star, and has no light of its own, so that even a large planet moving round a nearby star would be too faint to be seen through any telescope we have yet made.

The Sun is only one of a hundred thousand million stars making up what we call the Galaxy – and our telescopes can show us many millions of other galaxies, so that the number of stars in the universe is very great. It does not make sense to think that among all these stars, only our Sun has a family of planets. When we can build more powerful telescopes, and use them in space – beyond the top of the Earth's air – we may be able to find out.

CHAPTER 19

TRAVEL TO THE STARS

We have sent men to the Moon, and rockets out to most of the planets. Can we, then, hope to send rockets to the stars? The answer must be 'no', simply because it would take too long. A spacecraft such as *Voyager 2* could not reach even the nearest star in less than several thousands of years, and we would have no way of keeping in touch with it. In any case, we do not know which stars are likely to have families of planets.

Of course we have already sent up four spacecraft which will never come back – *Pioneers 10* and *11*, and *Voyagers 1* and *2* – but we will never know what happens to them after their signals become too faint for us to hear. If we are to reach another Solar System, we must have a spaceship much better and much faster than the Pioneers or the Voyagers.

One idea is to use nuclear power instead of our present-day rocket motors, but even this would mean a journey of hundreds of years. There is another difficulty, too. It seems that nothing can travel faster than light, and even if we could make a rocket that could work up to 186,000 miles (300,000 kilometres) per second it would still take years to reach any of the stars.

There have been many stories and many television programmes about interstellar travel (that is to say, travel to the stars). Most people will have watched the adventures of Dr Who in his *Tardis*, and the crew of the starship *Enterprise* in *Star Trek*. The trouble is that at the moment we have no idea of how journeys of this kind can really be made.

On the other hand, it is also true that travelling in a *Tardis* is no more unlikely to us than television would have seemed a hundred years ago. We will have to wait for some new discovery which will allow us to cross space without using rockets. When this will happen, we do not know; it may be quite soon, or it may never happen at all.

We must also ask whether there can be life on planets moving round other stars. There seems to be no life in the Solar System except on the Earth, but this is because the other planets in the

Artist's impression of the *Daedalus*, a starship for the future, first suggested by the British Interplanetary Society.

The *Bussard Ramjet*, another design for an interstellar craft which could one day explore beyond the Solar System.

Sun's family are too unfriendly; some have no air and others have poisonous atmospheres, while the giants such as Jupiter and Saturn do not even have solid surfaces. But if we could find a planet in another Solar System which was very like the Earth, we might expect life. It might be of the same kind as ours, though again we cannot be sure.

The only way in which we can hope to find out, at the moment, is to pick up radio signals from other intelligent beings. Radio waves move at the same speed as light, so that they could reach us from the nearest stars in only a few years. Efforts have already been made to listen out for messages from the stars, but up to now we have had no success.

Of course, everyone has heard stories about flying saucers or UFOs (Unidentified Flying Objects), but there is no reason to think that any UFOs are visiting spaceships. Other races, far away across the Galaxy, may be much more advanced than we are, and may even have learned how to travel between the stars, but there is no sign that any of them have yet come to see us!

One day, we may learn how to make interstellar travel possible, but until that happens we must be content with exploring our own part of the universe – the Solar System.

CHAPTER 20
WHAT NEXT?

Much has happened during the past thirty or forty years. What is likely to happen next – say before the year 2030?

More artificial satellites will be set up. The Russians already have one proper space station, *Mir*, and well before 2030 there will be others. By 2030 we must hope that all the countries of the world will be working together, so that the space stations will be international. The stations will be of great value in many ways – for weather forecasting, communications and so on.

One very exciting launch was that of the space telescope, named the Hubble Telescope in honour of the great American astronomer, Edwin Hubble, who died some years ago. The Hubble Telescope was put into a path 200 miles (500 kilometres) above the Earth's surface, and is able to see much further and much more clearly than any telescope on the ground, because it will not have to look through a thick layer of atmosphere.

Some people still ask: 'Why go into space, when there is so much to be done here on the Earth?' I hope that in this book I have been able to give some of the reasons. Yet we also know that rockets can be used in time of war. If there is another major war, we could easily wipe out all life on the Earth. Let us hope that nothing of the sort will happen.

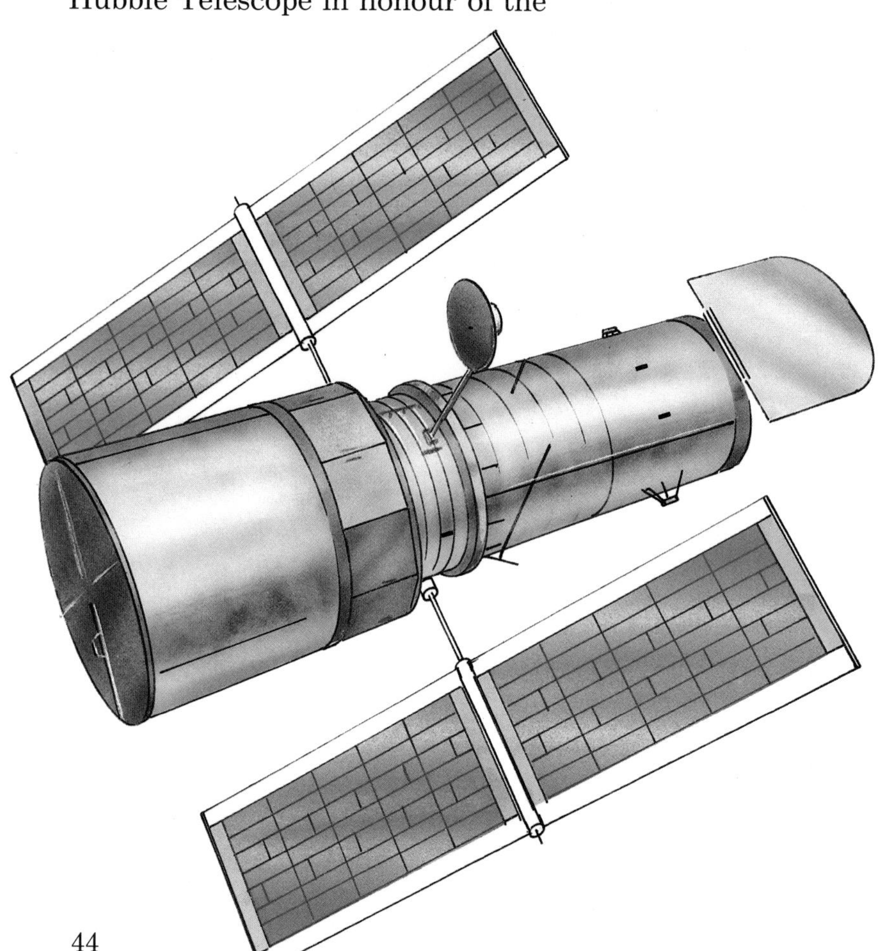

The Hubble Space Telescope will enable us to learn more about the universe than any earth-based telescope.

Artist's impression of a future manned mission to Titan, Saturn's largest moon.

If all goes well, men will go back to the Moon in a few years' time, and the Lunar Base could be built well before the year 2030 – perhaps even by 2000. Mars is our next target. The Martian Base could be set up before 2030, though there are still many problems which we have to solve before we can be sure. There will be new unmanned rockets to Venus, Mercury and the outer planets; we may hope that a spacecraft will go to Titan, Saturn's largest moon, to find out what lies below the orange clouds which hide its surface. We may even send a rocket to the lonely, icy little world of Pluto.

Holidays on the Moon may not be possible by 2030, but certainly it is likely that many people will have been there. Some of you now reading this book may well go into space – perhaps to a space station, perhaps even to the Lunar Base.

We are living in exciting times. There can be no doubt that the next few years will be more exciting still.

TEST YOUR SKILL!

I wonder how much you have learned by now? Most people enjoy testing themselves, so here is a quiz. You will find the answers on page 48.

1. Why can a rocket fly in space, but an aircraft cannot?
2. Name the first artificial satellite, launched by the Russians on 4 October 1957.
3. What was *Telstar*, and when was it launched?
4. Why is Yuri Gagarin so famous?
5. What is the Earth's escape velocity, in miles or kilometres per second?
6. Why is there a part of the Moon which we can never see from Earth – and when was it first photographed?
7. Who was the second man to walk on the Moon?
8. Why cannot we send a rocket from Earth to Mars by the shortest possible path?
9. Is Venus hotter or colder than the Earth?

10. Could you ever walk about on Mars, outside the Martian Base, without wearing a spacesuit? If not, why not?
11. Name the first spacecraft to pass by Jupiter.
12. Are there any seas on Mars?
13. Name Saturn's largest satellite, which has a dense atmosphere.
14. Have we yet sent a spacecraft to Pluto?
15. Name the unmanned spacecraft which passed through the head of Halley's Comet in 1986.
16. What is the colour of the sky as seen from Mars?
17. Which American astronomer is the NASA Space Telescope named after?
18. What is a light year?
19. If you go to the Moon, will you seem lighter or heavier than you do on Earth?
20. Name the three planets which were bypassed by *Voyager 2* between 1977 and 1987.

USEFUL DATES IN SPACE TRAVEL

1895	First scientific papers about space-travel (Tsiolkovskii).
1926	First liquid-fuel rocket launched (R. Goddard, USA).
1937–45	V.2 rockets developed by a German team led by Wernher von Braun.
1949	First step-rocket fired, from White Sands (USA).
1957	Launch of the first artificial satellite, *Sputnik 1* (USSR).
1958	Launch of the first American satellite, *Explorer 1*, and discovery of the Van Allen zones.
1959	First successful unmanned probes to the Moon: *Lunas 1, 2* and *3* (USSR). Pictures obtained of the Moon's far side.
1961	First manned space flight: Yuri Gagarin (USSR).
1962	First American orbital flight (John Glenn).* *Mariner 2* flew past Venus: first successful interplanetary mission (USA).* First television communications satellite (*Telstar*, USA).
1965	First 'space-walk' (Alexei Leonov, USSR).* First successful probe to Mars (*Mariner 4*, USA).
1966	First controlled landing on the Moon (*Luna 9*, USSR).* *Venera 3* (USSR) becomes the first probe to land on another planet: Venus (it crashed, and no data were received)
1967	First casualty in space; Vladimir Komarov killed on landing.
1968	First manned flight round the Moon (*Apollo 8*, USA).
1969	Armstrong and Aldrin (*Apollo 11*) land on the Moon.
1971–2	Detailed mapping of Mars, from *Mariner 9* (USA).
1973	*Pioneer 10* makes the first fly-by of Jupiter.* Launch of *Skylab*, America's first space-station.* First close-up pictures of Mercury (*Mariner 10*, USA).
1975	First pictures obtained from the surface of Venus (*Venera 9*, USSR).
1976	*Vikings 1* and *2* (USA) make controlled landings on Mars.
1977	*Voyagers 1* and *2* launched to the outer planets.
1979	Launch of the first Ariane, Europe's satellite launcher.* *Voyagers 1* and *2* fly past Jupiter.
1980	First good close-range views of Saturn, from *Voyager 1.*
1981	First launch of America's Space Shuttle.
1983	IRAS, the Infra-Red Astronomical Satellite, transmits valuable data for most of the year.
1986	*Voyager 2* obtains the first close-range views of Uranus.* Disaster with America's *Challenger* Shuttle; crew killed.* Five successful missions to Halley's Comet.
1988	Romanenko (USSR) completes the longest space-mission so far, on space-station *Mir.*
1989	*Voyager 2* scheduled to fly past Neptune.

ANSWERS

1. Because an aircraft cannot work unless there is air around it, while a rocket 'kicks against itself' and does not need surrounding air. **2.** *Sputnik 1.* **3.** The first television communications satellite. It was launched in 1965. **4.** He was the first man to go up in a spaceship (in April 1961). **5.** 7 miles (11 kilometres) per second. **6.** Because the Moon spins round in the same time that it takes to make one journey round the Earth (27.3 days). The first photographs of the far side were taken by *Luna 3* in 1959. **7.** Edwin ('Buzz') Aldrin. (Neil Armstrong was the first.) **8.** Because it would mean using more fuel than the spaceship could carry. **9.** Much hotter, because it is closer to the Sun, and its thick atmosphere shuts in the Sun's heat. **10.** No, because Mars does not have enough atmosphere. **11.** *Pioneer 10.* **12.** No – again because Mars does not have enough atmosphere. **13.** Titan. **14.** No. If Neptune is successfully bypassed by *Voyager 2* in 1989, Pluto will be the only planet we have not yet been able to study at close range. **15.** *Giotto.* **16.** Pink, because of the dust in Mars' atmosphere; the atmosphere itself is too thin to make the sky blue. **17.** Edwin Hubble. **18.** The distance traveled by a ray of light in one year; almost six million million miles (9.6 million million kilometres). **19.** Lighter. Because the Moon has a weak pull of gravity, you will have only one-sixth of your Earth weight. **20.** Jupiter, Saturn and Uranus.